EFFECTIVE NUCLEAR AND RADIATION REGULATORY SYSTEMS

PROCEEDINGS SERIES

EFFECTIVE NUCLEAR AND RADIATION REGULATORY SYSTEMS

WORKING TOGETHER TO ENCHANCE COOPERATION

PROCEEDINGS OF AN INTERNATIONAL CONFERENCE
ORGANIZED BY THE
INTERNATIONAL ATOMIC ENERGY AGENCY
IN COOPERATION WITH THE
EUROPEAN COMMISSION/JOINT RESEARCH CENTRE
AND HOSTED BY THE
GOVERNMENT OF THE NETHERLANDS
AND HELD IN THE HAGUE, 4–7 NOVEMBER 2019

INTERNATIONAL ATOMIC ENERGY AGENCY
VIENNA, 2022

COPYRIGHT NOTICE

© IAEA, 2022

Printed by the IAEA in Austria
November 2022
STI/PUB/2034

IAEA Library Cataloguing in Publication Data

Names: International Atomic Energy Agency.
Title: Effective nuclear and radiation regulatory systems : working together to enhance cooperation / International Atomic Energy Agency.
Description: Vienna : International Atomic Energy Agency, 2022. | Series: Proceedings series (International Atomic Energy Agency), ISSN 0074–1884 | Includes bibliographical references.
Identifiers: IAEAL 22-01545 | ISBN 978–92–0–143822–5 (paperback : alk. paper) | ISBN 978–92–0–143722–8 (pdf)
Subjects: LCSH: Nuclear facilities — Security measures — Congresses. | Nuclear facilities — Safety regulations — Congresses. | Nuclear energy — Law and legislation — Congresses. | Radioactive substances.
Classification: UDC 621.039.58 | STI/PUB/2034

FOREWORD

The 2019 International Conference on Effective Nuclear and Radiation Regulatory Systems was the fifth in a series of conferences on effective nuclear and radiation regulatory systems. The conference built on conclusions and deliberations of previous conferences in order to review issues and developments important to the global nuclear regulatory community and focus on their key role in ensuring safety and security.

Since the previous conference in 2016, there have been several significant events of relevance to nuclear and radiation regulation. The International Conference on Nuclear Security: Commitments and Actions was held in 2016. The ministerial declaration from the conference emphasized the importance of strong national legislative and regulatory frameworks for nuclear security. The Seventh Review Meeting of the Contracting Parties to the Convention on Nuclear Safety was held in 2017. A number of Contracting Parties reported on challenges faced by regulatory bodies, including the absence of legislation to provide adequate financial resources to enable the recruitment and retention of personnel to deliver an effective regulatory capability.

In 2017, the International Conference on Topical Issues in Nuclear Installation Safety: Safety Demonstration of Advanced Water-Cooled Nuclear Power Plants provided participants the opportunity to discuss, among other things, challenges to regulating non-conventional reactor types, such as modular high temperature gas reactors and molten salt reactors. The discussions emphasized the importance of a stable and well established regulatory framework for completing such nuclear projects.

The International Conference on the Physical Protection of Nuclear Material and Nuclear Facilities in 2017 highlighted the importance of the adoption of adequate legislation and regulations to implement the Convention on the Physical Protection of Nuclear Material and its 2005 Amendment, the use of IAEA Nuclear Security Series publications and international cooperation for nuclear security.

In 2018, the IAEA published Guidance on the Management of Disused Radioactive Sources, which supplements the Code of Conduct on the Safety and Security of Radioactive Sources. The Guidance aims to promote a more rigorous radiation safety and security culture. It also describes options for the management and protection of disused radioactive sources and outlines the responsibilities of relevant parties, including regulatory bodies. In particular, each State should ensure that legislation and regulations include provisions for the safe and secure management of disused radioactive sources.

The Contracting Parties to the Joint Convention on the Safety of Spent Fuel Management and on the Safety of Radioactive Waste Management convened their Sixth Review Meeting in 2018. Many Contracting Parties presented the most recent improvements to their legal and regulatory framework. The Contracting Parties also noted increasing the capacity of their regulatory bodies, improving licensing processes for disposal facilities, improving regulatory inspection programmes and taking measures to reinforce safety culture within the regulatory bodies.

The Third International Regulators Conference on Nuclear Security was held in 2019. The conference sought to further strengthen and sustain national, regional and international cooperation and enhance capacity building for nuclear security worldwide. The participants emphasized the importance of establishing and maintaining a comprehensive legislative and regulatory framework for nuclear security. The conference encouraged all countries to conduct a self-assessment of their nuclear security regimes based on the IAEA's International Physical Protection Advisory Service (IPPAS) Guidelines and to host IPPAS missions. The conference highlighted the importance of addressing the interface between safety and security and recommended the continued promotion of activities that address both areas in an integrated manner whenever justified.

This publication summarizes the International Conference on Effective Nuclear and Radiation Regulatory Systems which was attended by over 200 participants from 75 Member States and five international organizations in 2019. During the meeting, four keynote presentations and 35 invited papers were given throughout the opening session, five technical sessions and a special panel. Each day of the conference was accompanied by a panel discussion which allowed additional discussions as well as questions and answers. This publication includes the opening addresses, a summary of the conference and the conference President's summary and conclusions of the conference. The supplementary files, available on-line, contain the presentations and posters from the conference

The IAEA wishes to thank the contributors involved in the preparation of this publication. The IAEA officer responsible for this publication was S. Mallick of the Office of Safety and Security Coordination.

CONTENTS

OPENING SESSION

WELCOME ADDRESS

J.C. LENTIJO

Deputy Director General, Department of Nuclear Safety and Security
International Atomic Energy Agency,
Vienna

Mr President, ladies and gentlemen, distinguished guests, dear colleagues, good morning!

Welcome to the International Conference on Effective Nuclear and Radiation Regulatory Systems.

First, let me warmly thank the Government of the Netherlands for hosting this important conference. I also thank the European Commission for cooperation and support. And, last but not least, I thank my friend Carl-Magnus for serving as Conference President.

This conference is the fifth in a series of conferences on effective regulatory systems. The first was held 2006 in Moscow, with later conferences held in Cape Town, Ottawa and, most recently, in 2016 in Vienna. These conferences are important to keep the momentum in our joint efforts to strengthen regulatory systems. I know national representatives have been inspired to make improvements at home following earlier conferences, and I expect no less this time. The conferences and their recommendations for action also shape the Agency's work to help Member States strengthen regulatory systems. The president from the last conference, Mr Liu Hua, will speak later today to share what actions have been taken in response to the conclusions from the last conference. I encourage all of you to also share actions taken since the previous conference as this enables others to benefit from your experience.

Dear colleagues,

This Conference has attracted a high level of interest. Registrations increased by a quarter as compared to the last conference. This reflects the importance placed by Member States on the regulation of the nuclear and radiation safety and security.

The theme for this week is 'Working Together to Enhance Cooperation'. The conference aims to highlight how cooperation to address regulatory challenges improves regulatory effectiveness worldwide.

The programme includes four keynote presentations, eight topical sessions, a special panel discussion on emergency preparedness and response and poster sessions. It's an impressive, interesting programme and I regret that I my schedule does not allow me to participate beyond today.

Conference speakers will share their efforts to address ongoing challenges such as capacity building for regulatory infrastructure, knowledge management, and safety and security culture. They will also highlight emerging issues. These include those associated with new technologies, those related to ageing nuclear power plants and the back-fitting of current safety measures to existing nuclear power plants. Other emerging issues to be discussed include how to handle counterfeit and fraudulent items, decommissioning of nuclear installations, waste disposal, and the interface between safety and security. These topics are important cornerstones of nuclear safety and security, and I encourage you to take active part in the discussions.

Dear colleagues,

Let me now take a few moments to highlight some of the priorities in the Agency's latest Nuclear Safety Review and Nuclear Security Report that are relevant to the conference topic.

One priority listed in the documents is continued support to countries embarking on a nuclear power programme. This includes support to develop regulatory frameworks and to strengthen regulatory infrastructure.

Another priority is continued support for capacity building related to the regulation of small and modular reactors and research reactors.

We also prioritize supporting Member States in building capacity to strengthen their emergency preparedness and response arrangements. This includes assistance to Member States in the preparation, conduct and evaluation of emergency exercises.

Assisting Member States in the application of the Agency's safety standards is also a priority. This includes the International Basic Safety Standards, known as GSR Part 3. We assist Member States in the management of radioactive sources from cradle to grave.

In line with another priority, we will continue to assist States in strengthening their nuclear security regimes. We will also ensure that safety standards and nuclear security guidance alike consider the implications for both safety and security – this is what often is referred to as the safety–security interface. The peer review and advisory services we offer are part of our assistance to help Member States apply the standards and the guidance. These include services such as IRRS, AMRAS, IPPAS and EPREV.

Our peer review and advisory missions also are a key component of our work to foster international cooperation, which is the final priority I would like to mention. Though the regulation of safety and security are the responsibility of individual Member States, international cooperation helps all do better. I encourage all Member States to make full use of these services, conducted upon request.

To conclude, let me thank the conference Programme Committee and the IAEA Secretariat staff for their work related to the conference. I wish you success for the coming days. I look forward to learning about the outcome of your discussions in the President's Report.

Thank you.

WELCOME ADDRESS

C.-M. LARSSON
Conference President
Australian Radiation Protection and Nuclear Safety Agency (ARPANSA)
Australia

Ladies and Gentlemen, may I call the meeting to order. Thank you.

Representatives of the International Atomic Energy Agency, the Government of the Netherlands, the European Commission, and conference participants, welcome to the Sixth IAEA Conference on Effective Regulatory Systems.

I am Carl-Magnus Larsson, Chief Executive Officer of the Australian Radiation Protection and Nuclear Safety Agency, ARPANSA, and I have been asked by the IAEA to be the President of this Conference. It will be my honour and pleasure to guide you through our deliberations over the next few days.

I will shortly provide some more information on how this Conference will run. But before that I want to introduce my distinguished colleagues here on the podium; Mr Juan Carlos Lentijo, Deputy Director General, Safety and Security of the IAEA, Ms Maria Betti, Director Nuclear Safety and Security of the European Commission, Mr Jan van den Heuvel, Chairman of the Authority for Nuclear Safety and Radiation Protection of the Netherlands, and Mr Hua Liu, Vice Minister of Ministry Ecology and Environment and Administrator of the National Nuclear Safety Administration of China and President of the 2016 Conference on Effective Regulatory Systems.

Also here is Mr Shahid Mallick, Scientific Secretary of this Conference, who has worked tirelessly with the Programme Committee and with the IAEA Team to make this event happen.

Let me also, as Conference President, again extend my welcome to all participants. I would also like to congratulate the Programme Committee and the Conference Secretariat for their work, which has now resulted in very good turnout, with about 250 participants, well above previous Conferences. We also have twice the number of posters compared to previous events and I am sure that, against this backdrop, we will have very interesting few days ahead of us.

We have a programme with subject matter areas that span across nuclear and radiation safety, and with cross-cutting themes and overlapping areas such as the interface between safety and security.

We all know, when we talk about safety, radiation protection, physical protection, security of assets and information, emergency preparedness and response, or any other aspect of practices that involve radiation, we actually talk about people.

Education, training, recruitment, awareness, leadership and management for safety, and communication are all essential elements, and they all sit with people. Ageing facilities and ageing workforce are challenges but at the same time there are also new technologies and new applications, medical use of radiation for diagnosis or radiotherapy being one prime example. Those challenges all have to be addressed by people. This Conference will deal with all of these aspects, and in a holistic manner that takes the technical, managerial, organisational and behavioural factors into account.

Also, the success of this Conference depends on interaction between people, being us, the participants. We will encourage this interaction by polling questions, to for example, seek your views on issues and priorities. At the end of each day, we will have time for panel discussions involving the Chairs of the sessions of the day. Jointly, and with the help of the Secretariat we will try to inject questions into these discussions, and tease out the main issues, conclusions and potential recommendations. The first panel, at the end of today, will also include the keynote speakers.

I will have 15 minutes in the morning of the following day to summarise the major points and in that manner, we can build the President's Report from the Conference, as we go.

I remind you that the theme for this Conference is "Working together to enhance cooperation", which of course is one of the manners by which we can holistically establish effective regulatory systems.

Thank you.

SESSION SUMMARIES

SESSION SUMMARIES

CONFERENCE OBJECTIVES

The objective of the conference was to share regulatory experiences related to improving the effectiveness of nuclear and radiation regulatory systems, addressing the international framework for the safety and security of nuclear and other radioactive material. The focus of the conference was on how to work together to address cross-cutting regulatory areas. Among the expected outcomes of the conference were:

— Enhanced international cooperation to support embarking countries;
— Strengthened regulatory interfaces between nuclear and radiation safety and nuclear security;
— Improved regulatory effectiveness through the application of a graded approach and the use of regulatory experience;
— Improved anticipation and management of cross-cutting regulatory areas considering regulatory lessons learned from other industries;
— Identified strategies and actions for the future, including topics for consideration by governments and regulatory bodies, which includes interfacing with technical support organizations and international organizations.

The conference included an opening session, five technical sessions, three plenary panel discussion, and one topical panel.

OPENING SESSION

The conference was opened by IAEA Deputy Director General Mr. Juan-Carlos Lentijo. Mr. Lentijo thanked the Government of the Netherlands for hosting the event and the European Commission for their cooperation and support in organizing the conference. He emphasized the importance of this series of conferences in the ongoing efforts to strengthen regulatory systems. He noted that this conference aimed to highlight how cooperation to address regulatory challenges can improve regulatory effectiveness worldwide. Mr. Lentijo encouraged the delegates to share their actions taken since the previous conference held in Vienna, Austria in 2016, to enable others to benefit from their experience.

Mr. Lentijo noted some of the ongoing challenges faced by regulatory bodies, such as capacity building for regulatory infrastructure, knowledge management and strengthening safety and security culture. He highlighted several emerging issues to be discussed during the conference, including new and innovative technologies, ageing of nuclear power plants and the back-fitting of current safety measures to existing nuclear power plants. In addition, counterfeit and fraudulent items, decommissioning of nuclear installations, waste disposal, and the interface between safety and security were also topics to be addressed in the programme.

Mr. Lentijo then highlighted some of the Agency's priorities of relevance to the conference, including the continued support to Member States to develop their regulatory frameworks, strengthen their regulatory infrastructure and strengthen their emergency preparedness and response (EPR) arrangements. Mr. Lentijo noted that assisting Member States in the application of the Agency's safety standards remains a high priority. He emphasized the importance of the Agency's peer review and advisory missions such as the Integrated Regulatory Review Service (IRRS) which are a key component of fostering international cooperation. Mr. Lentijo stated that the Agency will continue to assist Member States in the management of radioactive sources from cradle to grave and will continue to assist States in strengthening their nuclear security regimes. The Agency will also ensure that the safety standards and nuclear security guidance will consider the interface between safety and security. Mr. Lentijo noted that the conference and its outputs will shape the Agency's future work to continue to supporting Member States in strengthening their regulatory systems.

Mr. Jan van den Heuvel, Chair of the Netherlands Authority for Nuclear Safety and Radiation Protection, welcomed delegates to the conference. He stated that he was pleased that the Netherlands were invited to host this international conference and noted the large number of participants from many countries around the world.

Mr. Jan van den Heuvel highlighted one challenge that was facing many regulatory bodies around the world, namely the oversight of radioactive material in and out of regulatory control. He noted the particular challenge arising from orphan radioactive sources. While this is a global issue, Mr. Jan van den Heuvel remarked that it is particularly relevant to the Netherlands, which is a large importer of scrap metal.

Mr. Jan van den Heuvel emphasized the value of this international conference in facilitating the exchange of experience and information. However, the ultimate benefit to be obtained from this conference will be the application of the knowledge and experience shared here this week into activities in the field.

Ms. Maria Betti welcomed the delegates on behalf of the European Commission (EC). She stated that fostering a global partnership of nuclear regulators to ensure the highest standards of nuclear and radiation safety and nuclear security is a key objective which benefit all stakeholders. She noted the importance of this objective in the context of mitigating climate change and accelerating the energy transition towards decarbonisation.

Ms. Betti considered that international cooperation of regulators should be strengthened to promote an international culture for safety and security. Standards should be harmonized, mutual recognition by regulatory authorities ensured, embarking countries supported and regulatory interfaces between nuclear and radiation safety and security strengthened.

She highlighted the fora for cooperation between European regulatory bodies such as the Western European Nuclear Regulators Association (WENRA), the European Nuclear Safety Regulators Group (ENSREG), and the European Nuclear Security Regulators Association (ENSRA). Ms. Betti stated that the European Union is strongly engaged in fostering regional and international cooperation on nuclear regulation.

In his opening remarks, the Conference President Mr. Carl-Magnus Larsson, emphasized the importance of the human dimension in, among others, nuclear safety, radiation protection and emergency preparedness and response.

Mr. Larsson remarked that, while the ageing of facilities and of the workforce are ongoing challenges for regulatory bodies, there are also challenges to be faced from emerging and innovative technologies and new applications using existing technology. He noted that this conference will deal with these aspects in a holistic manner, considering the technical, managerial, organisational and behavioural factors. Mr. Larsson encouraged the delegates to interact, seek views and exchange ideas on the issues to be discussed and identify priorities for the future.

Mr. Liu Hua, the President of the Vienna Conference, provided an update of activities and progress made since 2016. The theme for the previous conference was 'Sustaining Improvements Globally'. The conference proposed issues for consideration for international cooperation, for governments and for regulatory bodies.

Mr. Liu highlighted progress with these issues including those relating to international cooperation, namely:

— Improving the interface between nuclear safety and nuclear security;
— Encouraging greater participation in the Convention on Nuclear Safety (CNS) and the Joint Convention on the Safety of Spent Fuel Management and on the Safety of Radioactive Waste Management (Joint Convention);
— Strengthening the IAEA peer review services.

Mr. Liu noted that, in response to these international cooperation issues, the IAEA convened a Technical Meeting on the safety–security interface in 2018. The number of Contracting Parties to the CNS and the Joint Convention had increased by 9 and 10 respectively since 2016. The internal IAEA Peer Review and Advisory Services Committee had been created and 37 IRRS missions had been carried out since 2016, indicating the value placed on this service by Member States.

Mr. Liu summarized the actions taken in China since 2016, including the implementation of the Nuclear Safety Law, the creation of two separate Divisions in the regulatory body (one for experience feedback and one for nuclear safety coordination), and the development of an integrated management system for the regulatory body.

In his keynote speech, Mr. Petteri Tiippana summarized the first of the European Union Topical Peer Reviews to be conducted. The theme of the first review was 'Ageing management of nuclear power plants and research reactors'. This review, conducted from 2017 to 2018, was the most important safety-related exercise after the post-Fukushima stress tests in Europe. The main outcome of the review was that ageing management programmes exist for nuclear power plants (NPPs) in all countries reviewed; they conform with the IAEA safety standards and WENRA Safety Reference Levels and no major deficiencies were identified. However, for research reactors, ageing management programmes are neither regulated nor implemented as systematically and comprehensively as they are for NPPs. Therefore, ageing management for research reactors requires further attention from regulators and licensees.

Some of the challenges identified by the peer review process included:

— Further development of performance indicators is needed to enable consistent evaluation of the effectiveness of aging management programmes.
— Research and development (R&D) for non-invasive inspection methods is necessary for detection of local corrosion issues.

— Objective and comprehensive acceptance criteria for ageing management of concrete structures is needed.

Mr. Tiippana concluded that this peer review process provided an opportunity to identify areas of good practice and areas for improvement along with identifying common issues and learning from each other. He considered the ongoing process of topical peer reviews will be an excellent instrument to ensure and enhance nuclear safety.

In her keynote presentation, Ms Kristine Svinicki informed the conference that the United States Nuclear Regulatory Commission (USNRC) has cautiously moved from a prescriptive regulatory approach to a risk-informed, performance-based approach. This move allowed a better focus of attention on design and operational issues commensurate with their importance to public health and safety. Ms Svinicki noted that USNRC staff can make decisions more efficiently through the increased use of risk insights to determine and guide the quality and level of effort appropriate for a given regulated activity. The approach of increasing leverage from risk insights has been instrumental as the USNRC adapts to dealing with new technologies such as small modular reactors and other advanced NPP designs.

Ms Svinicki explained that the USNRC continues to address expected human capital changes in its organization and to enhance its workforce. The USNRC is committed to ensuring that there is an appropriate organizational culture, an expert staff, and the processes and tools necessary to continue to accomplish its safety and security mission.

Mr. Christer Viktorsson summarized the challenges faced by regulatory bodies in countries embarking or considering on a nuclear power programme, the 'newcomer countries'. He noted that a skilled workforce for a nuclear power programme is needed for a long period of time. He emphasized the importance of an early assessment of the development needs for the regulatory framework and for establishing an independent regulatory body. Mr. Viktorsson stated that, while international support is important for the embarking countries, this support must be tailored to meet each country's individual needs. He recognized that newcomer countries may need to rely on foreign support at least in early stages of a programme, but building national capacity is vitally important. Mr. Viktorsson highlighted the benefit from cooperation and support from the vendor country regulatory body and the support provided by the IAEA. He noted that the United Arab Emirates (UAE) had hosted 11 IAEA peer review missions across all areas, including IRRS missions. In addition, the UAE had hosted the first IAEA Integrated Nuclear Infrastructure Review (INIR) Phase 3 mission.

Mr. Viktorsson stated that the UAE Federal Authority for Nuclear Regulation (FANR) was set up in 2009 with a mandate for regulating nuclear safety, security and safeguards as well as radiation generators and sources. He emphasized that it is important that the development of the regulatory infrastructure keeps pace with the NPP programme. In a period of 10 years, the UAE had built a regulatory framework that facilitated the transition from initial licensing of an NPP to operation. He noted that infrastructure development in the UAE is continuing with, for example, the inauguration of a state-of-the-art emergency centre and a nuclear R&D centre in the very near future.

In his keynote speech, Mr. Khammar Mrabit described the capacity building experience of the Moroccan Agency for Nuclear and Radiological Safety and Security (AMSSNuR) in upgrading the national regulatory framework. Mr. Mrabit noted the efforts to reinforce the nuclear safety and nuclear security infrastructure and developing capacity building and improving safety and security culture. He presented the AMSSNuR programme of continuous improvement to the regulatory infrastructure through hosting international peer reviews such as the IRRS, the International Physical Protection Advisory Service (IPPAS) and the Emergency Preparedness Review (EPREV) Service and implementing their results.

Mr. Mrabit stated that a systematic approach to capacity building has been used with the 'four pillars' of education and training, human resource development, knowledge management and knowledge networks. He emphasized the importance of cooperation at the national, regional and international levels with multilateral and bilateral agreements. Mr. Mrabit noted that, in developing a national strategy for nuclear safety and security education and training, it is important to establish what is needed, what support is available, and what programme(s) will meet the need.

SESSIONS 1 AND 2: REGULATING NUCLEAR INSTALLATIONS

These two sessions addressed some of the current challenges associated with the regulation of nuclear installations. Such challenges include regulation of aged NPPs and back-fitting safety standards to existing NPPs; dealing with counterfeit and fraudulent items (CFI); regulation of innovative nuclear technology; the application of a graded approach to regulatory oversight; the management of legacy sites; the regulation of disposal facilities and the lessons from decommissioning.

The following points were noted during these sessions:

— The operating lifetime of NPPs continues to increase, resulting in changes to the mechanical and physical properties of structures, systems and components important to safety. This issue of physical ageing is becoming more important. In addition, the issue of obsolescence of electrical and mechanical components needs to be addressed at the older NPPs. In some Member States, the legal framework has been adapted to accommodate efforts to extend the operating lifetime of NPPs. The changes include requirements for an NPP to be judged against the latest regulatory standards and be subject to specific inspections for ageing.

— The back-fitting requirements in some Member States follow a regulatory philosophy that allows NPPs to continue operation as long as they are deemed to be safe. However, continuous improvement is expected to bring them as close as reasonably possible to current expectations for safety. The design and configuration of an NPP after back-fitting can be very different to those at the start of operation, for example through back-fitting safety systems against the principles of the Vienna Declaration on Nuclear Safety for the prevention of accidents and mitigation of their radiological consequences.

— The established regulatory oversight arrangements for NPPs have not been able to rule out the presence of CFI. This has resulted in arrangements being adapted to take account of good practices of other non-nuclear regulators that have encountered similar problems, for example the drug safety regulatory bodies.

— The CFI issue is considered to have arisen from weaknesses in safety culture and, as a result, may be difficult to detect. In one Member State, the regulatory body has developed a web page to encourage anonymous reporting (whistleblowing) and one CFI issue was reported at a medical installation.

— The graded approach to regulation is a process where the regulatory activities are commensurate with the risks and characteristics of a facility or activity. The graded approach to regulation can be applied when:
  - Safety requirements are met;
  - Safety margins are sufficient;
  - Defence in depth (DiD) has been maintained.

The application of a graded approach to regulatory oversight improves the flexibility of a nuclear regulatory system so that it can adapt and respond to the challenges such as from innovative technologies.

— The peak in decommissioning activities and the associated waste management considerations will continue to generate problems for regulatory bodies, including site remediation and licensing of disposal facilities. The need to develop appropriate regulations for decommissioning is becoming more important and a graded approach should be applied.

— Nuclear decommissioning can become more efficient and economical by following good practices and principles. This can also strengthen public trust and the confidence of other stakeholders. Continuous knowledge management and cooperation is vital and sharing and retaining local site knowledge was considered essential.

— Extensive challenges exist to regulating uranium legacy sites. Issues to be addressed include land erosion, seismic activity, potential for flooding and landslides. Effective regulatory oversight of these legacy sites and their environmental remediation is necessary to ensure land can eventually become available for future use. International cooperation is helping to overcome these challenges by providing technical assistance through, for example, support provided by the IAEA in improving the qualification of regulators and operators at these sites.

— The disposal of spent nuclear fuel presents considerable challenges to regulatory oversight, as almost no prior experience or examples exist worldwide. Pursuing the disposal option is a learning process for the regulator and the operator. This requires flexibility of approach and persistence of both parties. It is not possible to set comprehensive safety standards at the beginning of the process. It is vital to build trust between the regulator and the operator as well as with the public and the international community. The key factors to success are an active competent regulator with commitment and courage.

— The future operation of a disposal facility for spent nuclear fuel will also require a new regulatory approach and new skills. These skills will have to be maintained during the long operating lifetime projected for disposal facilities.

— Regulatory bodies need to continue to demonstrate that regulatory decisions are proportionate to the risk being regulated for protecting the public and the environment. This is particularly important when the risk posed by a facility changes significantly, as is the case in the transition period from operation to decommissioning of NPPs. Building public trust in the risk-informed approach to regulation is essential.

— New and innovative technologies being used at, for example, nuclear research facilities give rise to regulatory challenges. Regulators need to respond to these challenges and provide solutions that will ultimately strengthen the regulatory framework. One example of an innovative technology posing regulatory challenges was the use of nuclear fuel in an aqueous solution, to implement innovative technology for production of medical radioisotopes. The challenges to be addressed included the absence of physical barriers such as a fuel matrix and fuel cladding. These challenges were addressed using a graded approach to regulation, along with the development of new regulations and improving the licensing and inspection procedures.

SPECIAL PANEL – EMERGENCY PREPAREDNESS AND RESPONSE

The Special Panel on emergency preparedness and response (EPR) noted that significant progress had been made in strengthening EPR by Member States and international organizations. The Panel acknowledged the efforts of the IAEA to support Member States in their implementation of IAEA Safety Standards Series No. GSR Part 7, *Preparedness and Response for a Nuclear or Radiological Emergency*, and in harmonizing their national EPR arrangements. The exceptional collaborative effort involved in the preparation of GSR Part 7 was highlighted, involving the co-sponsorship of 13 International Organizations. The Panel provided the opportunity to discuss a number of issues, including the importance of conducting emergency exercises, communicating with decision makers and the public, achieving an appropriate balance between the benefit of introducing protective measures and the harm that may result, and the need to further strengthen cooperation and collaboration nationally and internationally.

The Special Panel noted the following points:

— The IAEA support to Member States in harmonizing national EPR arrangements was recognized, as well as the efforts of other international bodies, including the Heads of European Radiological Protection Competent Authorities (HERCA) and WENRA.
— The importance of cross-border cooperation in EPR was highlighted, particularly where some NPPs were sited close to national borders. However, there is still more work to be done in harmonizing arrangements. A consistent response to a nuclear accident is necessary where neighbouring countries should adopt and implement harmonized arrangements, particularly the arrangements to protect the public and the environment from the harmful effects of ionizing radiation.
— The importance of communicating with the two key stakeholders in EPR, the public and the decision makers, was emphasized. It is essential that these stakeholders understand why protective measures are being taken. However, it was recognized that there may be a tendency to take stronger protective measures than can be justified from a purely radiation protection or science-based perspective.

— The 'benefit–harm' balance was identified as an issue to be addressed, namely, achieving an appropriate balance between the benefit of introducing protective measures and the harm that may result from their implementation. It was considered that, if protective actions, such as evacuation, are undertaken using criteria at low dose levels, this can result in more harm than good.

— The psycho-social consequences of unnecessary evacuation and the non-radiological impact of prolonged evacuation and relocation need to be better understood. These consequences need to be considered in EPR arrangements where they may have a strong non-radiological impact on public health.

— The Panel discussed the key role in EPR of the nuclear safety and radiation protection regulatory authorities. Although recognizing the official role of these authorities in several Member States is limited to giving advice on radiation protection measures, there was a view that they could take the lead in coordination and communication.

— The importance of conducting emergency exercises was emphasized, including tabletop exercises to test national and international EPR arrangements to ensure they are in line with IAEA safety standards. The national exercises could be extended to allow participation of authorities from other Member States which could serve to strengthen regional and international cooperation and collaboration.

— The number of national bodies with responsibilities for EPR was recognized. Within any Member State, such bodies include, but are not limited to, those with responsibilities for civil protection, transport, law enforcement and health. It was recognized that it is not always easy to achieve effective collaboration with all the authorities involved. In some Member States, instances of poor communication between the authorities and a lack of appreciation of the different roles and responsibilities are issues to be addressed.

— To ensure that the overall national EPR arrangements are appropriate, all relevant organisations need to be effective. There is a need for good communication among all bodies involved in EPR. A well prepared regulatory body for nuclear and radiation safety is a start, but on its own is not sufficient.

— The 'move or remain' balance is the key to decision making for evacuation in the case of severe accident management. However, it is likely that decisions will be made when there are large uncertainties, but such decisions should always be underpinned by science and informed by experience.

— The usefulness of generic criteria for protective actions with numerical ranges was questioned, in particular where there could be differences in the interpretation and application of these criteria between countries. This could lead to psycho-social consequences as a result of a value being deemed 'safe' in one country and 'unsafe' in another country. The potential for confusion by the public was clearly identified by the Panel. The importance of differentiation and understanding of differences between application of reference levels, generic criteria and operational criteria was stressed.

SESSIONS 3 AND 4: REGULATING RADIATION AND MEDICAL FACILITIES

These sessions covered the challenges in regulating activities and sources in new applications. The session also covered the efforts to establish an appropriate regulatory infrastructure for radiation safety and the necessary staff competencies of regulatory bodies. Presentations were given on the safety and security regulation of radioactive sources, the detection and prevention of orphaned sources, and the interaction of multiple regulatory bodies.

The following points were noted during these sessions:

— Keeping pace with rapid developments in technology is a challenge for regulatory bodies. Advances in technology take place much faster than the speed with which regulations can be introduced. An appropriate regulatory response to quickly changing technology could be the use of lower level guidance which may be easier and quicker to produce.
— Regulatory challenges can also arise from the use of existing technologies in different ways, particularly when these technologies are used outside of the original design intent. Regulatory bodies may not always be aware of these changes of use.
— Research facilities can pose regulatory challenges from the non-routine nature of the work, the non-standard design of some facilities, and the limited availability of operational feedback.
— Medical applications are associated with a wide range of risks ranging from negligible to significant. The application of the graded approach is being applied to these applications based on many decades of regulatory experience. Using the graded approach, the regulatory activities are proportional to the hazard and the number of such activities being undertaken.
— The regulatory environment for medical uses of radiation is complex and can involve numerous organizations, including radiation regulators, health authorities, product safety authorities, professional societies and colleges. In addition, there are challenges in countries that have a federal constitution, where there can be many pieces of legislation for radiation protection, that operate simultaneously and not entirely consistently when it comes to implementation.
— The overall regulatory framework can be strengthened by effective inter-agency coordination and cooperation through formal and informal agreements.
— There are developments in small-field radiotherapy and particle therapy where patient safety and treatment success depend on the right dose in the right place and the margin for error is very small. International collaboration is a major pathway for achieving international equity in treatment outcomes and patient safety.
— Occupational safety in some of the medical uses of radiation also requires attention, particularly in the preparation of high activity unsealed radioactive sources. While the half-life of these sources is generally very short, significant occupational radiation doses can be incurred, particularly in the production of radiopharmaceuticals.
— Preventing the loss of control of radioactive sources is a significant regulatory issue which is exacerbated by the large number of radioactive sources in circulation around the world. The use of tracking systems for each radioactive source was discussed and considered to present opportunities for the future to strengthen the regulatory control of sources.

— The detection of orphan radioactive sources is also an ongoing issue faced by regulators worldwide. Recent experience on dealing with the discovery and recovery of such sources has shown the benefits to be realized from international cooperation and learning from each other.

— Suggestions were made during this session regarding resurrecting the proposed Code of Conduct on *Control of Transboundary Movement of Radioactive Material Inadvertently Incorporated into Scrap Metal and Semi-finished Products of the Metal Recycling Industries*, which was drafted in 2014.

SESSIONS 5 AND 6: CROSS-CUTTING REGULATORY AREAS

These sessions addressed several cross-cutting regulatory areas, including the safety–security interface, nuclear knowledge management and networks, managing, capturing and using regulatory experience to improve effectiveness, capacity building, the importance of international collaboration through R&D, and the Technical and Scientific Support Organization (TSO) Forum. The importance of public communication and involvement and the enhancement of public awareness in regulatory activities were also considered.

The following points were noted during these sessions:

— In some countries, the regulation of safety and security is undertaken by the same organization. Regulatory assessments and inspections are performed considering both areas and are coordinated from the beginning. It was considered that it is more effective and efficient to manage safety and security and its interface from a single body as opposed to a diverse set of organizations.

— Measuring and demonstrating regulatory effectiveness is a complex and challenging issue. A systematic approach to capturing regulatory experience can provide evidence-based assurance on the status of the regulatory environment. A regulatory assurance framework can identify successful practices and behaviours and areas for improvement. The framework can be useful to support regulatory interactions with the licensees and other stakeholders. Measuring and reporting on regulatory effectiveness can be used to influence future planning and developing and adapting regulatory strategies. However, this requires the commitment of sufficient resources to be effective.

— Experience feedback from internal and external sources is essential to the effectiveness of a regulatory body. International partnerships with established regulatory bodies provide an important framework to learn lessons and strengthen regulatory processes and capabilities. This experience can also be applied to the regulatory oversight for new and emerging technologies.

— Regulatory bodies in nuclear power embarking countries should gather information at the earliest possible stage to strengthen their framework. Resources should be dedicated to assessing the strengths and deficiencies and apply the lessons to improve regulatory effectiveness. The relationship with the regulator of a vendor country is extremely valuable for the transfer the requisite knowledge and experience.

— External independent observations and assessments, such as through the IAEA services of IRRS and IPPAS, are extremely valuable in strengthening the regulatory framework and improving the effectiveness of regulatory oversight.

— Enhancing communication with the public and increasing their awareness of regulatory activities is important. Some Member States publish inspection results and safety and security guides and actively seek feedback on regulatory activities. This allows a wide range of perspectives to be considered on regulatory activities. The value of advisory councils was also highlighted as an independent means of addressing important issues related to regulatory oversight.

— Education and training for a regulatory body can be a slow and steady process, which may require many years before the workforce can effectively fulfil the regulatory mandate. A structured approach to national capacity building is necessary so that it can be planned and executed at a pace in parallel to its operational requirements while maintaining a focus on sustainability.

— Developing a comprehensive human resource management strategy is key to meeting both the short and long term development goals and objectives for a regulatory body. Competencies should be linked to the overall regulatory strategy, with a focus on technical, managerial and soft skills. Training programmes should be phased to increase the complexity of skills and competencies to match the needs and goals of the national nuclear programme as well as the demographics of the human resources.

— There is a need for appropriate technical and scientific support to strengthen the effectiveness of national regulatory bodies. Some national TSO capability is necessary, and this capability may be internal or external to the regulatory body. The TSO Forum promotes collaboration among all TSOs and organizations interested in developing and strengthening technical and scientific capacity.

— There was a widely shared concern that developing and maintaining the TSO capability is not always adequately resourced by Member States. There is a need to strengthen cooperation among TSOs and improve their capabilities to provide advice to regulatory bodies. The recently published IAEA TECDOC No. 1835 *Technical and Scientific Support Organizations Providing Support to Regulatory Functions* was highlighted that describes the characteristics and functions of a TSO. In order to assess the capabilities and gaps of Member States TSOs, national workshops will be offered to improve their strategies and optimize their efforts.

— Research and development activities are important to develop the knowledge and skills, tools and methods to support regulatory oversight. For research facilities to be maintained, there needs to be better international coordination noting that these facilities are essential to strengthening global nuclear safety.

— The knowledge needs of regulators are becoming much broader and more complex and coordinating knowledge development and transfer is a challenge. Member States have the responsibility for national knowledge management. Regional bodies such as the European Commission act to coordinate the management of knowledge as well as its dissemination. Effective knowledge management requires international cooperation.

SESSION 7: LEADERSHIP AND MANAGEMENT FOR SAFETY AND SECURITY

This session addressed safety culture and security culture within regulatory bodies. It also examined the regulatory oversight of programmes established to strengthen human performance to achieve a high level of safety and security and establishing integrated management systems. Some presenters provided an overview of their approach to regulatory oversight of safety culture as well as efforts to promote, maintain, and improve safety culture in their regulatory bodies. The achievements, difficulties and challenges in this area were summarized, including initial and follow-up safety culture assessments and organizational change.

The following points were noted during these sessions:

— Incidents related to technical factors have reduced significantly over time, while those related to human behaviour have gained more prominence. Consequently, just addressing technical factors is not enough to achieve a continuous improvement in safety. Safety is achieved through humans, organizations and technology interacting together. Providing an appropriate balance to human, organizational and technical factors is key to accident prevention and continuous improvements in safety.

— A key issue when developing and implementing an integrated safety management system was a systemic approach to considering human, organizational and technical factors. A management system that addresses these factors holistically reduces the probability of latent errors and supports their early identification and timely correction. A good balance is needed between the structure of the management system and the culture of the organization.

— There can be a tendency within organizations to develop more written procedures while the same objective could be achieved through training programmes, skills development and seminars. There cannot be rules and procedures for everything, and an appropriate organizational culture will foster proper management, for example the balance between the policy documents, instructions and procedures that guide an organization and the shared norms, values and assumptions that influence its behaviour.

— The culture of a regulatory body is owned at the executive level and flows all the way down the organization. Regulators have an important role in ensuring the establishment of proper safety management system and its effective implementation. A strong safety culture in the regulatory body can strengthen the safety culture across the regulated industry.

— The incorporation of innovation into regulatory activities and adapting to changes in the operating environment is an important issue in improving regulatory activities. Regulators need to innovate and remain current to reflect the changes in the industry they are regulating. Future forecasting of challenges and resource needs can provide adequate preparation for organizational changes.

— When adapting to a changing operating environment, such as adopting a different regulatory approach or dealing with innovative technologies, a successful transformation process is a key aspect in maintaining an effective regulatory body. Such a process concentrates mainly human related issues in terms of skills, training, and leadership behaviour.

— To achieve success in transforming a regulatory body, it is important to communicate widely, internally and externally, on why and how the organization is changing. Consequently, regulators should not be isolated from other stakeholders but be responsive to feedback. Fully utilizing an organization's human asset is essential to continue to meet its regulatory mission.

— In some Member States, the nuclear licensees have been performing self-assessment of safety culture for many years. Recently published regulatory documents establish requirements and provide guidance for licensees to foster safety and security culture and conduct periodic safety and security culture assessments.

— Some nuclear regulators have completed and reported on regulatory safety culture self-assessment against international principles of safety culture. Their ongoing efforts to promote, maintain and improve regulatory culture for safety include staff surveys, safety culture working groups, a knowledge management initiative, and processes to deal with differences of professional opinion.

— Some non-nuclear regulators have conducted safety culture assessments resulting in improved and updated regulatory practices. The value of a joint Ibero-American Forum of Radiological and Nuclear Regulatory Agencies (FORO) – IAEA project on safety culture was highlighted. An organisational learning tool was described where lessons were learned from radiological events and disseminated to authorized facilities and other stakeholders.

— Communicating regulatory processes and decisions, and monitoring progress were considered to be very valuable exercises in addressing safety culture, in particular conducting regular surveys and communicating the results within the regulator and with the licensees. This can identify areas of improvement and enhance the culture of the regulator and of the licensees.

— Plans for regulatory process improvements based on the results of surveys are key to addressing deficiencies. International cooperation has proved to be valuable in enhancing safety culture and sharing best practices and lessons learned. Further improvements to safety culture face challenges such as resistance to change and complacency.

SESSION 8: STRENGTHENING INTERNATIONAL COOPERATION

The session began with Mr. Gustavo Caruso presenting the role of the IAEA in strengthening international cooperation for nuclear and radiation safety and nuclear security. He noted that, while both safety and nuclear security are national responsibilities, the IAEA plays a central role in promoting international cooperation to assist Member States in their efforts to fulfil these responsibilities. This central role has been reaffirmed by Ministerial Declarations at international conferences on both nuclear safety and nuclear security. He explained that the IAEA actively cooperates with a wide range of other international organizations including EC, the Nuclear Energy Agency of the Organisation for Economic Co-operation and Development (OECD/NEA), the World Health Organization (WHO), the World Association of

Nuclear Operators (WANO), and the International Criminal Police Organization (INTERPOL).

Mr. Caruso stated that the IAEA offers nearly 20 Peer Review and Advisory Services to Member States. Requests for these services continues to increase and every year many missions are conducted across all safety areas. Of particular interest to this conference were the EPREV, Advisory Mission on Regulatory Infrastructure for Radiation Safety (AMRAS) and IRRS services. He reiterated an earlier comment made by Mr. Liu Hua that 37 IRRS missions had been carried out since 2016.

Mr. Caruso remarked that, to help IAEA implement its activities, it collaborates with designated Member State centres that focus on research, development and training for nuclear safety and security. These collaborating centres include those in Argentina, Costa Rica, Hungary, Japan, Republic of Korea, Mexico and Spain. The benefits to Member States from IAEA recognition of these centres include sharing resources with others, addressing issues of common interest and networking. Mr. Caruso noted that these centres help Member States in their efforts toward achieving the targets identified in the UN Sustainable Development Goals.

Mr. Caruso highlighted some recent trends in the need for IAEA support. Around 75–80% of Member States receiving IAEA support still need additional support to develop a national regulatory infrastructure consistent with the Agency's safety standards. In response, the Agency has reviewed its approach to the provision of assistance and developed the concept of the Consolidated Plan for Safety (CPS). The aim of the CPS is to provide consistent, coherent and result oriented assistance to Member States.

Mr. Caruso highlighted the importance of international cooperation against the backdrop of the nuclear accidents at Chernobyl and Fukushima Daiichi. He noted that international cooperation has become a hallmark of nuclear safety, resulting in innumerable peer reviews being performed by IAEA and others, bilateral and multilateral assistance efforts, the safety conventions, and the globally recognized IAEA safety standards.

Mr. Caruso closed his presentation by noting that the IAEA has an important role to play in supporting Member States in the attainment of the UN climate change targets and Sustainable Development Goals. The IAEA does this by providing guidance and assistance for deploying safe, secure and safeguarded nuclear technology.

Panel Discussion

The panel members provided their views and experience on international cooperation.

Ms. Anna Bradford (USA) outlined the aims and activities of the Small Modular Reactor (SMR) Regulators' Forum and its contribution to international cooperation. The Forum was formed in 2015 to understand member's regulatory views on common issues and to capture good practices and understand key challenges that are emerging in SMR regulatory discussions. Ms. Bradford remarked that the Forum addressed three issues for both light-water and non-light-water SMR designs:

i.) The application of the graded approach and clarifying the regulatory view of grading and what it means in practice;

ii.) The smaller size of emergency planning zones (EPZs) being proposed by some SMR vendors and the current practices and strategies for understanding how flexible EPZs are established in order to have a common position on this issue;

iii.) The different approaches of SMR designers addressing DiD and attempted to develop common positions around certain regulatory practices to ensure that the fundamental principles of DiD are maintained.

Mr. Hans Wanner (Switzerland) highlighted the activities of WENRA. He noted the comprehensive IAEA platforms that exist for global and regional cooperation. He stated that the accidents at Chernobyl and Fukushima Daiichi both highlighted the key role of cultural issues in safety. Mr. Wanner considered that the importance to safety of human and organizational factors is increasing but measuring and assessing 'culture' is not easy. At a national level, he noted the emphasis on the social skills of the workforce becoming more important, particularly how these skills impact the cultural behaviour. He remarked that the IAEA peer review services such as the IRRS contain a mix of expertise from different countries and cultures with different thought processes. He considered that this is a benefit that should be further exploited and views on cultural issues should be exchanged at the international level.

Mr. Alfredo de los Reyes (Spain) summarized the relevant international activities of the FORO in promoting and maintaining high levels of radiation protection, nuclear safety and nuclear security in its member countries. He explained that FORO's technical programme has focused on priority thematic areas carried out through an IAEA extra-budgetary programme, funded by voluntary contributions from FORO members. These arrangements strengthen the resolve to continue working together, to share a harmonized vision on key issues and to cooperate with the IAEA and other organizations to disseminate the results. Mr. de los Reyes emphasized the importance of FORO not overlapping with the work of the IAEA. He provided examples of the FORO work, including the stress tests for NPPs, control of inadvertent radioactive material in scrap metal and recycling industries, and emergency preparedness and response.

Mr. Ghislain Pascal (European Commission) highlighted the efforts of the EC in collaborating with the IAEA. He emphasized the importance of ensuring the efforts of the EC complement those of IAEA and not overlap. He noted that the European Union appreciates the benefit of the IAEA's peer review services such as IRRS and ARTEMIS, which were a valuable tool for to fulfil their legal obligations on nuclear safety and waste management. Mr. Pascal stated that the EC continues to support a variety of IAEA activities, such as the Regulatory Cooperation Forum and the joint efforts for environmental remediation in Central Asia. He explained that the EC and IAEA share the aims of promoting effective nuclear safety culture and implementation of the highest nuclear safety and radiation protection standards, and the continuous improvement of nuclear safety.

Mr. Philip Webster (Canada) described the IAEA Regulatory Infrastructure Development Project (RIDP) which aims to help countries strengthen their national regulatory infrastructure. The RIDP addresses both radiation safety and security in a harmonized manner that is tailored to the needs of individual Member States. The project activities involve different IAEA resources, including expert and advisory missions and national and regional training events. The activities under the RIDP can offer support to the participating countries in areas related to the national policy for safety and security, the associated regulatory framework, authorization and inspection processes, physical protection and management systems. Mr. Webster noted that the

RIDP complements assistance provided by the IAEA through national and regional technical cooperation projects to strengthen regulatory infrastructure. For example, the project helps introduce or strengthen procedures and systems to safely and securely handle and control radioactive sources used in medicine, industry and research.

Mr. Bismark Tyobeka (South Africa) highlighted the activities of the Regulatory Cooperation Forum (RCF) and the important contribution it makes in supporting newcomer countries. The RCF is a regulator-to-regulator forum promoting collaboration and cooperation among Member States. It brings together countries with well-established nuclear power programmes (the donors), with those countries considering the introduction or expansion of nuclear power programmes (the recipients). The IAEA provides the secretariat to the RCF. Mr. Tyobeka noted that the RCF was very appreciative of the goodwill of the donor countries in providing support. He underlined the importance of international cooperation to South Africa and, in particular, the support provided by the RCF to the national regulatory body. Regarding other aspects of international experience for newcomer countries, Mr. Tyobeka stated that hosting an IRRS mission was essential to evaluate and strengthen the national regulatory framework. He also emphasized the benefits of becoming a Contracting Party to the IAEA safety conventions.

CLOSING SESSION

CLOSING REMARKS

The Conference President, Mr. Carl-Magnus Larsson, presented his summary and conclusions of the conference. These are presented separately below.

In his closing remarks, Mr. Michael Huebel (European Commission) noted the importance of international cooperation and collaboration. The EU stress tests and peer reviews provided good examples of cooperation but more needs to be done. The conference highlighted the importance of maintaining regulatory effectiveness and it also provided an opportunity to learning from other sectors. He hoped that this conference would prove to be of benefit to regulators around the world.

Mr. Marco Brugmans (Netherlands) stated it was an honour for the Government of the Netherlands to host this international conference and a personal privilege to be involved in the event. He thanked the EC and IAEA for their cooperation and collaboration.

Mr. Caruso (IAEA) thanked the delegates for their participation in the conference. He expressed his appreciation to the President for his excellent summary and noted that it will constitute an excellent guide for future activities in strengthening regulatory effectiveness.

Mr. Caruso reiterated the remarks of Deputy Director General Lentijo in his opening address and emphasized the important benefits to be gained by Member States from the peer review and other services offered by the IAEA. He noted the continuous increase in demand for these services and stated that the IAEA continues to stand ready to assist Member States in their efforts to strengthen international cooperation. Mr. Caruso thanked the Government of the Netherlands for hosting the conference and the Joint Research Centre of the European Commission for their invaluable cooperation. He once again thanked Carl-Magnus Larsson for his sound stewardship as Conference President and the members of the organising committee. He expressed his appreciation to the panellists, speakers, chairpersons and poster presenters and the delegates for their very active and essential contribution to the success of the conference.

Mr. Caruso then declared the conference closed.

PRESIDENT'S SUMMARY AND CONCLUSIONS OF THE CONFERENCE

C.-M. LARSSON
Conference President

The President of the conference, Mr. Carl-Magnus Larsson (Australia), noted that the conference had attracted 238 registered participants from 75 Member States and 5 international organizations. A total of 69 poster presentations had been submitted. The number of participants was on par with, or surpassed, the number of attendees at previous conferences in this series. This illustrates the need for regulators to meet periodically to exchange information on experiences and approaches that help the regulatory community to improve the effectiveness of nuclear and radiation regulation, for the purpose of protection of people and the environment.

The President had identified a number of themes that had been highlighted throughout the conference and referred to in detail in previous sections of this President's Report. The President summarized these themes as follows:

Old and new nuclear facilities

The conference provided an opportunity to discuss challenges associated with ageing nuclear facilities. Significant focus has been placed on ageing management of power reactors and it was noted that ageing management should also be consistently applied to the large number of ageing research reactors in the world, as well as to other facilities that utilize nuclear technology for, e.g., isotope production. In the coming years, an increasing number of power and research reactors will reach their end of life and the number of major decommissioning projects will increase. This places further emphasis on the need for safe and publicly acceptable solutions for the back-end management of the nuclear fuel cycle, including decommissioning, management of radioactive waste and used nuclear fuel, disposal, and remediation of sites.

In contrast, a number of countries are newcomers to nuclear power, and have provided a number of good illustrations of how an effective regulatory framework can be cautiously and carefully established from a low baseline, building on international experience and best practice. Other newcomer countries as well as established nuclear countries can learn from these examples. However, the back-end management of radioactive waste and spent fuel may not yet have been fully considered and require further attention. In addition, new nuclear technologies are at various stages of development, ranging from the drawing board to construction. Knowledge management will be key to success, either it relates to ageing existing facilities, new facilities in newcomer countries and in countries with an established nuclear programme, or to the implementation of new nuclear technologies.

Large accidents

The possibility of a large nuclear accident can never be ruled out. IAEA provides significant support to its Member States in implementing the requirements of

25

GSR Part 7, prepared in an exceptional international effort involving 13 co-sponsoring organizations. The requirements support national and cross-border plans and arrangements for harmonized emergency preparedness and response systems.

While international collaboration can provide both solutions and guidance in this regard, important issues remain that can only be resolved within the national cultural context. This includes communication on value-laden terms such as "risk" and "safe", and the need to understand and properly address the psychosocial impact of fear and involuntary evacuation and/or relocation in an emergency. All aspects of health, including the physical, mental and social aspects of well-being must be considered, but will not alleviate the psycho-social impact if trust in the responsible authorities is not there, or if messaging from authorities is un-coordinated or even conflicting.

'Numbers' such as reference levels can guide decision making but make no sense if not properly communicated to the public, emergency workers, volunteers and the public. The confusion that can arise when attempting to understand, or even access relevant information in an emergency. This matter was clearly illustrated by the Dutch writer and performer, known under the artist name TINKEBELL, who has extensively visited the Fukushima Prefecture and shared her experiences and views in a well-attended side event to the conference.

Sources

Maintaining regulatory control of radioactive sources, either still in use or disused, is a key preventive measure to avoid radiation accidents caused by improper management, loss, theft or acts with malicious intent. At the same time, it is a significant challenge, considering the number of sources in circulation or stored. Innovative measures for source tracking look promising. International co-operation is of the essence and there may be scope for strengthening or expanding the agreements and Codes of Conduct that are already in place.

Manufacturing of sources for nuclear applications, in particular unsealed sources for use in medicine and research, often involves manual handling of small samples with extremely high activity concentrations. Skin contamination with minute quantities of product can cause severe radiation injury in very short time. Recent events in several countries suggest that increased regulatory attention should be given to the safety of such manufacturing facilities, and the use of the sources for various purposes including their administration to patients undergoing diagnosis or treatment.

Radiation in medicine

Medical applications have a wide range of risks associated with them, ranging from negligible to significant. The regulatory environment for medical uses of radiation is complex. It involves numerous organizations, including radiation regulators (sometimes multiple regulators in countries with a federal constitution), product safety authorities, health ministries, insurance systems, professional societies and colleges, and others.

It is not reasonable to expect that international collaboration can overcome all problems related to such fragmentation, as they have their origin in the constitution and in national arrangements for health care. Nevertheless, international collaboration can promote the safety of the patients and the workers and can support safe introduction of diagnostic techniques and radiation therapy in countries with a less developed infrastructure for safety, where the population has justifiable expectations of improved health care. New technologies are introduced in radiation therapy, examples are small-field radiation therapy and increased investment in particle therapy to improve effectiveness of treatment and, by reducing the 'collateral' damage of surrounding tissue, significantly improve the treatment experience of the patient. International collaboration is an important mechanism for promoting safety in medical radiation applications, and international equity in health outcomes.

Culture

Human factors are a recognized element of safety. While structures and processes can guide safety measures, they cannot solve all safety-related issues. A culture for safety (and security) that is supported and implemented in practice by management (leading by example) assists staff at all levels to do the right thing with the safety objective in mind.

However, improving safety culture faces obstacles such as resistance to change, attitudes including taking safety for granted and lack of situational awareness, and production pressures. It is important that regulators lead the way and seek to first understand their own safety culture, and in doing so improve their understanding of their licensees' safety culture. Safety culture is a shared trait, and, in fact, many regulators are also operators of facilities and sources licensed under legislation that also govern their regulatory activities. Regulators are increasingly paying attention to their own safety culture but also to the effectiveness of their regulatory activities – do the regulatory actions promote a cultural shift among regulated entities that leads to the desired safety outcome? Do the regulatory actions adopt a graded approach so that requirements on operators are commensurate with the hazard and risk? And, is the 'regulatory burden' distributed in a risk-informed manner that is understood by the regulated entities? Again, international collaboration and sharing of experience can assist countries to improve the safety of their workforce, public and environment, regardless of the size of their radiation and/or nuclear programmes.

The safety–security interface is an area where many parties have agreed for considerable time that improved integration is desirable, while retaining the unique characteristics of both safety and security. The integration should be considered throughout the lifecycle, starting with 'safety and security by design', and include arrangements for emergency management. Importantly, it should be supported by a culture that considers both safety and security.

Capacity building

Capacity building entails four elements, being: education and training; human resource (workforce planning) development; knowledge networks; and knowledge management.

The problems associated with capacity building are many, including 'information overload', harmonization, competing priorities, generational shifts, uncontrollable changes in national policies and priorities… the list can go on and on. The introduction of new and innovative technologies, either in nuclear technology or in other areas such as medicine, industry and research, requires recruitment of staff with appropriate skills, upskilling of existing staff and recurrent training. Maintaining and building capacity is an issue not only for the regulator and operator but must be supported by the national educational system and, at the international level, through regulatory and operating experience feedback and in collaboration on research and development. The need for capacity building is shared between established nuclear power countries, newcomers to nuclear power, countries with advanced health and non-nuclear industry sectors, and countries where these sectors still require development. IAEA provides important platforms by which capacity building can be sustained and supported, in international collaboration.

Challenges and issues faced by regulatory bodies to be addressed in international cooperation

The President considered that the identified themes pose challenges that can be addressed by regulatory bodies in international collaboration. The President also proposed that the next conference in this series can provide a mechanism for follow-up on progress made in relation to these challenges. The challenges identified were:

— Capacity building enabling the regulatory body to respond to emerging and innovative technologies in nuclear, medical, R&D and other applications of nuclear and radiation technologies. The IAEA should continue supporting Member States, especially embarking countries, in building their regulatory infrastructure through education and training, knowledge management and human resource development activities. This will include supporting Member States through the Consolidated Plan for Safety (CPS). This challenge is shared with the theme 'Issues for consideration by Governments'.
— The need for strengthening international cooperation through networks such as the Global Nuclear Safety and Security Network (GNSSN), while at the same time avoiding duplication and overlap.
— The development and implementation of the concept of the graded approach to regulation, including toward aspects such as risk, culture and safety performance. This includes adapting the management system to allow for efficient and effective implementation of a graded approach.
— Developing and implementing programmes for using regulatory experience and evaluating and monitoring regulatory effectiveness, building on international cooperation and experience and suited to the size and nature of the national programme for nuclear and radiation facilities. In this regard, the IRRS was considered a good service for strengthening the regulatory framework and IAEA should continue providing peer review and advisory services focusing on regulatory bodies.
— Maintaining and enhancing transparency and openness when engaging with the public and other stakeholders. Communicate in a manner (for example during emergencies) that earns trust and confidence in the regulator among all stakeholders, including the public.
— The increasing need to decommission major nuclear facilities will increase the pressure on the back-end management of the nuclear fuel cycle and on establishment of policies, plans and enabling legislation. While this is well known, progress is not universal. This challenge is shared with the theme 'Issues for consideration by Governments'.
— Establishment and enhancement of regulatory approaches within the management system that fully consider the common culture of the regulated entities as well as the regulatory bodies; this includes consideration of a 'one culture' organizational approach for safety and security, and for extending

consideration of human, organizational and technical interfaces to non-power nuclear applications and radiation facilities.

— Ageing management of power reactors is receiving significant attention and with generally good outcomes but requires continued evaluation and monitoring. Ageing management considerations should be extended to non-power nuclear applications including ageing research reactors, radiopharmaceuticals production facilities, and storage facilities.

— Management of disused radioactive sources and those out of regulatory control and prevention of loss of regulatory control through preventive measures including tracking, registering and cross-border cooperation.

— International collaboration for optimizing research and development activities.

— Strengthening leadership and management for safety and security with emphasis on the young generation of regulators and practitioners.

Issues for consideration by governments

It is recognized that the regulatory body does not act in isolation and is dependent on the infrastructure developed by governments. The following issues were identified for consideration by governments, with the input and assistance from the regulatory body:

— Ensure the coordination of activities of all national bodies with interfacing and overlapping regulatory responsibilities, including for emergency preparedness and response.

— Develop mechanisms and frameworks for appropriate consideration of psycho-societal aspects associated with emergency actions, taking experience from non-nuclear and radiological emergencies into account.

— Promote national programmes for capacity building including education and safety infrastructure for nuclear power countries, for embarking countries and for countries using non-power nuclear applications, to ensure the availability of resources and infrastructure in line with the national needs.

Conclusions

The President concluded that:

— Since the previous conference, held in Vienna, Austria, in 2016 many improvements have been made to nuclear and radiation regulatory systems, many developments have taken place in countries that are bringing their regulatory infrastructure in line with international best practice, and many regulators have invested much effort in reflection on their culture and whether they are truly effective. But many challenges remain the same, some of them outside of the regulatory body's control.

— Sharing experience and lessons learned is key to sustaining improvements globally, but even more important is to understand the future and configure ourselves not based on problems we faced in the past but to deal with problems we will face in the next decade(s).

— The summary, actions and conclusions propose issues for consideration by regulators and governments. The intention is to stimulate us as regulators to take appropriate actions to respond to these issues and to update our regulatory colleagues on progress at the next conference on effective regulatory systems, to be held in three years' time. A willing Member State will be sought by the Secretariat to host the next conference in the series.

ANNEX

SUPPLEMENTARY FILES

The following presentations from this conference can be found on the publication's individual web page at www.iaea.org/publications. Please note that the Special Panel was a panel discussion and therefore no presentations were given.

SESSION 1: REGULATING NUCLEAR INSTALLATIONS

Switzerland back-fitting standards to existing NPPs
G. Schwarz

Detecting, preventing and dealing with counterfeit and fraudulent items (CFI): adapting the oversight processes
C. Quintin

Safety regulation of innovative nuclear facilities in the Russian Federation: challenges and solutions
A. Ferapontov

Safety regulations for aged reactors
M. Yasui

SESSION 2: REGULATING NUCLEAR INSTALLATIONS CONT'D

Innovation in Regulation – Enhancing Regulatory Effectiveness
R. Jammal

Management of uranium legacy sites in Uzbekistan: recent international activity
B. Kuldjanov

Regulating disposal facilities
J. Heinonen

Regulatory lessons to be drawn from industry experience in managing material and waste from nuclear decommissioning
C. Sanders

SESSION 3: REGULATING RADIATION AND MEDICAL FACILITIES

Graded approach applied to medical applications in France: from strategy to practical implementation
P. Chaumet-Riffaud

The detection and prevention of orphaned sources
M. Korse

Regulatory aspects of radioisotope production
S. Carvalho

Regulating Radiation Sources and Medical Facilities in the United States including challenges with new technologies and applications
J. Elee

Regulating medical and research activities using ion technology
P.K. Dash Sharma

Competencies of the Staff of Regulatory Bodies in Medical and Industrial Radiological Applications
M. Emacora

SESSION 4: REGULATING RADIATION AND MEDICAL FACILITIES CONT'D

Rostechnadzor's interaction on radiation safety and nuclear security with other regulators in the Russian Federation to enhance regulatory capabilities
E. Kudryaytsev

Regulatory Challenges for Reactor Based Nuclear Medicine Production
J. Scott

Establishing Radiation Safety Regulatory Infrastructure
S. Getachew

Regulating Radiation Sources and Medical Facilities
A. Nader

Radiation Safety Regulation of Nuclear Technology Utilization in China
Y. Jiang

SESSION 5: CROSS-CUTTING REGULATORY AREAS

Safety-security interface in the Finnish regulatory framework
M. Tuomainen

Challenges and Opportunities in Nuclear Knowledge Management and Networks
F. Wastin

ONR's Regulatory Assurance Framework - Driving Continuous Improvement in our Regulation
C. Tait

Technical and Scientific Support Organization (TSO) Forum – Supporting the Development of Technical and Scientific Capacities in Member States
G. Lamarre

Capturing and Using Regulatory Experience to Improve Effectiveness
N. Mughal

SESSION 6: CROSS-CUTTING REGULATORY AREAS CONT'D

Building capacity and capability United Arab Emirates nuclear regulator's perspective
S. Al Mansoori

UK progress in developing a new framework for measuring and reporting on regulatory effectiveness
M. Foy

Public communication and involvement – enhancement of awareness in regulatory activities in the Russian Federation
D. Bokov

BAPETEN Human Resource Development
L. Hakim

International efforts on R&D dedicated to safety expertise and decision making
P. Bueso

SESSION 7: LEADERSHIP AND MANAGEMENT FOR SAFETY AND SECURITY

Significance of Human and Organizational Factors in Ensuring Safety
D.K. Shukla

Graded Approach to Integrated Management Systems and Safety Culture: The Balance between Structure and Culture
A. Franzen

Nuclear Regulatory Transformation Activities in the United States
M. Doane

Safety Culture – The Canadian Nuclear Regulator`s Perspective
G. Lamarre

Internal Safety Culture at the Cuban Regulatory body: achievements, difficulties and challenges
R. Ferro

SESSION 8: STRENGTHENING INTERNATIONAL COOPERATION

Role of the IAEA in Strengthening International cooperation for nuclear and radiation safety and nuclear security
G. Caruso

POSTER SESSIONS

Topic 1: Regulating Nuclear Installations

Regulation of new reactor technologies in AERB and associated challenges
U. Chikkanagoudar

Regulation of nuclear fuel cycle facilities safety in the Russian Federation. Back end issues
A. Lavrinovich

Regulating Nuclear Installations in Vietnam
T.T. Tran

Software Failure Analysis using a Soft Computing Technique for the Reactor Protection System
H. Sallam

Development of Regulatory Infrastructure for Nuclear Installations in Ghana
E. Ampomah-Amoako

Safety Critical Software Failure Prevention Using Defence-in-Depth Approach
E. Shafei

Nuclear and Radiological Regulatory Commission (NRRC) In Saudi Arabia
M. Alharbi

Topic 2: Regulating Radiation and Medical Facilities

The Role of the Regulatory Authority in the Security of Radioactive Materials in Ghana
O. Agbenorku

Improving radiation-hygiene regulations in light of protection of the public from new nuclear activities: experience of the Republic of Belarus
L. Rozdyalouskaya

Establishing Regulatory Infrastructure For Radiation Safety: Sustainability to Enhance Safety and Security
S.B. Utami

Improvement of Indonesian government regulation on the licensing of radiation generator and radioactive material utilization
A. Hayani

Regulatory System on Radiation in Nepal: Long Overdue
K.P. Adhikari

Graded approaches for the authorization of ionizing radiation sources in Cameroon
J.F. Beyala Ateba

Evolution of Brazilian radiotherapy licensing in the last 5 years
C. Salata

Safety-Security interface in Madagascar: Challenges and Opportunities with the Implementation of the Code of Conduct on the Safety and Security of Radioactive Sources
J.L. Zafimajato

The Challenge of NORM Authorization in Indonesia
E. Yuliati

Strengthening Effectiveness of Regulatory System in Myanmar-Promoting and Maintaining Regulatory Infrastructure for the Control of Radiation Sources
K. Pa Pa Tun

Radiation Personnel Awareness towards Regulation in Nuclear Security based on Self-Assessment survey by IAEA
A. Fakhruddin

Establishing Regulatory Infrastructure for Radiation Safety in Malawi
C. Gamulani

Current Status and Challenges in the Development of Regulation for the Safety and Security of Transport of Radioactive Material in Indonesia
H.P. Yuwana

Challenges in regulating radiation sources in mining facilities subject to dam break scenarios
M.L. De Lara Costa

Regulating Medical and Industrial Applications of Radiation in Ghana: Challenges, Good Practices and Experiences
P.K. Gyekye

Evaluation of operation and quality control of mammography
C. Yegros

The preliminary Probabilistic Safety Assessment study of the CNSTN Gamma Irradiation Facility
W. Dridi

The Evolution of the Regulatory Infrastructure for Radiation Safety in Nigeria
O. Okoya

The Philippine Experience in Regulating Safety and Security of Category 1 and 2 Radioactive Sources
J.R. Fernandez

Strengthening Regulatory Control of Radioactive Sources: The Case of Tanzania
S. Sawe

Safety and security of radioactive waste
A. Chilulu

Regulatory requirements for building radiopharmacy facilities
L.A.M. Quiroz

Practice-specific challenges in the management of regulatory functions of radiation sources and medical facilities
N. Ramamoorthy

Regulating medical application of radiation in Bolivia
S. Ibanez Bravo

Correction Methods Applied on the Image Contrast of MPI in SPECT/CT Hybrid Systems: The Diagnostic Needs and the Necessary Regulations for Good Practice
N. Helal

Regulatory perspective and challenges of occupational radiological protection in medical practice
R.H. Alvarez

Establishment and development of standard operating procedures (sops) in diagnostic and therapeutic nuclear medicine services in Malaysia: coalition between regulators and medical institutions
S.I. Saufi

Regulatory Challenges in Regulating Radiation Sources in Medical use in Senegal
M.S. Tall

Artificial Intelligence Development on Proposed Methodology for Standardization on Evaluating Radiological Protection System Implementation in Regulatory Inspections: One Researching Agenda
B. Costa Filgueiras

Topic 3: Cross-Cutting Regulatory Areas

Enhancing the management of regulatory experience towards improving the regulatory process
D. Senior

Managing regulatory experience as part of continuous improvement at the Finnish Radiation and Nuclear Safety Authority (STUK)
M. Andersen

Public Communication: from Requirement to Reality
G. Gorashchenkova

Public Communication, Awareness, Involvement and Participation by PNRA
N. Mughal

Role of Nuclear Education Programs in the Enhancement of Nuclear Regulatory System in Turkey
B.B. Acar

The Nuclear Security Inspection Guideline for Thailand BNCT
R. Maneechayangkoon

Safety Culture in the Regulatory Body
I. Grlicarev

An Integrated Approach for Safety and Cyber Security of Digital I&C Systems in Nuclear Power Plants Combing Bowtie and Cyber Design Basis Threat Techniques
Md. Dulal Hossain

Improve nuclear safety with information technology in China
H. Peng

A comprehensive and integrated regulation of radiation protection, nuclear safety and nuclear security – The Swedish approach
M. Gustavsson

Towards effective regulatory infrastructure for control of radiation sources in Tanzania
J.E. Ngaile

Non-Nuclear Regulatory Experience
Z. Simic

E-licensing system for improving the effectiveness of Regulatory Functions
A. Al Remeithi

Establishment of a comprehensive Radiation Safety infrastructure in the UAE
B. Al Ameri

Integrating Regulatory Experience at the Canadian Nuclear Safety Commission (CNSC)
J.C. Poirier

Topic 4: Leadership and Management for Safety and Security

Implementing a RB Safety Culture Self-assessment through a Safety Culture Maturity Matrix
B. Bernard

Assessing safety culture within nuclear installations: Insights from a "Safety Culture Observation Process"
B. Bernard

Implementation of Integrated Management System in Atomic Energy Regulatory Board, India
S. Sinha

Safety Culture Self-Assessment at PNRA and PNRA response
R. Hammad

Integrated management systems for nuclear regulatory functions in Mexico
A. Nunez-Carrera

State management on radiation and nuclear safety
H. Nguyen

Experience in Implementing Nuclear Safety and Security Regulatory Processes by Regulatory Authority Information System (RAIS)
A. Simo

Sri Lanka's preparedness for nuclear and radiological emergencies
P.K. Koralalage

Nuclear Safety Culture in China
W. Guo

Regulator Approach to Promote the Safety Culture on Radiopharmacy Facilities
L.A.M. Quiroz

Inclusive Management of radiation safety and security: Involvement of people with special needs
M.F. Haris

An inclusive approach to the development of an effective ionizing regulatory environment in Jamaica
T.A. Warner

Topic 5: Strengthening International Cooperation

FNRBA's Contribution for Strengthening Regulatory Infrastructure in African Embarking Countries
A. Simo

International cooperation efforts of the nuclear regulatory authority, Ghana
E. Ampomah-Amoako

Support Requirements for Safety Management for Decommissioning of Research Reactors
N. Helal

Identifying and addressing support needs in Relation to Radioactive Waste management in Egypt
M. Abdel Geleel

Towards a European System of Nuclear Regulation: Enhancing Regional Cooperation in the Nuclear Field
K.T. Olajos

Lessons Learned: The Nuclear Safety Training and Tutoring Programme supporting Nuclear Regulators
Y. Stockmann

Regulatory experiences related to improving the effectiveness of nuclear and radiation regulatory systems
P. Ottavainen

Impact of Peer Review Missions on Regulatory Activities in Kenya
I. Mundia

Advances in the systems to exchange international data and information in Europe in case of radiological/nuclear emergency
M. de Cort

Strengthening Regulatory and Technical Capabilities through International Cooperation
A. Al Khadouri

CHAIRPERSONS OF SESSIONS

Session 1	S. GHOSE	Bangladesh
	S. CADET-MERCIER	France
Session 2	G. CHAI	China
	M. DOANE	United States of America
Special Panel	S.A. BHARDWAJ	India
	M. BRUGMANS	Netherlands
Session 3	F. OLLITE	Mauritius
	B.F. NDEYE ARAME	Senegal
Session 4	A. FACURE	Brazil
	N. RAMAMORTHY	India
Session 5	R. JAMMAL	Canada
	I. SOKOLOVA	Russian Federation
Session 6	M. GARRIBBA	European Commission
	M. ZIAKOVA	Slovakia
Session 7	Z.A. BAIG	Pakistan
Session 8	K. MRABIT	Morocco
	S.M. PETRICK CASAGRANDE	Peru

PRESIDENT AND VICE-PRESIDENT OF THE CONFERENCE

C-M. LARSSON
M. BRUGMANS

PROGRAMME COMMITTEE MEMBERS

C-M. LARSSON (President)	Australia
M. BRUGMANS (Vice-President)	Netherlands
O. LUGOVSKAYA	Belarus
A. FACURE	Brazil
P. WEBSTER	Canada
J. YU	China
M. HUEBEL	European Commission
V. RANGUELOVA	European Commission
K. ALM-LYTZ	Finland
F. JOUREAU	France
D.K. SHUKLA	India
N. ICHII	Japan
K. MRABIT	Morocco
I. SOKOLOVA	Russian Federation
B. TYOBEKA	South Africa
H. NILSSON	Switzerland
E. BENNER	United States of America
S. MALLICK	IAEA
E. BEAUPRE	IAEA
R. PACHECO JIMENEZ	IAEA
D. SENIOR	IAEA
S. HARVEY	IAEA
A. HALLOCK	IAEA

SECRETARIAT OF THE CONFERENCE

S. MALLICK	Scientific Secretary
M. KHAELSS	Conference Coordinator
E. BEAUPRE	Scientific Support
P. SALES BARBOSA	Administrative Support

ORDERING LOCALLY

IAEA priced publications may be purchased from the sources listed below or from major local booksellers.

Orders for unpriced publications should be made directly to the IAEA. The contact details are given at the end of this list.

NORTH AMERICA

Bernan / Rowman & Littlefield

15250 NBN Way, Blue Ridge Summit, PA 17214, USA

Telephone: +1 800 462 6420 • Fax: +1 800 338 4550

Email: orders@rowman.com • Web site: www.rowman.com/bernan

REST OF WORLD

Please contact your preferred local supplier, or our lead distributor:

Eurospan Group

Gray's Inn House

127 Clerkenwell Road

London EC1R 5DB

United Kingdom

Trade orders and enquiries:

Telephone: +44 (0)176 760 4972 • Fax: +44 (0)176 760 1640

Email: eurospan@turpin-distribution.com

Individual orders:

www.eurospanbookstore.com/iaea

For further information:

Telephone: +44 (0)207 240 0856 • Fax: +44 (0)207 379 0609

Email: info@eurospangroup.com • Web site: www.eurospangroup.com

Orders for both priced and unpriced publications may be addressed directly to:

Marketing and Sales Unit

International Atomic Energy Agency

Vienna International Centre, PO Box 100, 1400 Vienna, Austria

Telephone: +43 1 2600 22529 or 22530 • Fax: +43 1 26007 22529

Email: sales.publications@iaea.org • Web site: www.iaea.org/publications